The **Good** AND THE **Beautiful**

The STORY of INVENTION

Written by David Wiseman
Design by Kayla Ellingsworth

© 2022 Jenny Phillips
goodandbeautiful.com

Are you ready to become an inventor?

Maybe you have already designed a useful gadget or perhaps you have simply imagined new possibilities, but as children of God, each of us is born with the power to create. As you read about how fifteen amazing inventions came to be, try to think of ways you can improve the world through your own creative spirit!

The Nail:

Around 3400 BC, the ancient Egyptians really nailed it with their new idea to use pointed strips of bronze to bind things together. Other ancient civilizations, such as the Romans, also depended upon the nail to construct powerful empires. Before the 1790s, with the rise of machines to cut nails from raw iron sheets, individual nails were typically crafted by hand. The world now produces trillions of nails for a variety of purposes, but their basic design has been the same for thousands of years. From houses to horseshoes and boats to railways, nails connect much of the modern world—literally! The mighty nail proves that good things really can come in small packages.

Inventor tip:

An idea is never too small to have a big impact.

"In our private pursuits, it is a great advantage that every honest employment is deemed honorable. I am myself a nail-maker." —Thomas Jefferson

Fun facts:

- Due to the unique designs of nails over time, dating nails by their shape and material is one of the primary ways experts determine the age of antique furniture and historic buildings.

- Carpenters use the term penny to measure the length of nails. This name comes from medieval England when the price of nails was set by how many pennies it would cost to buy 100 nails.

- Before machines started mass-producing nails, shortages made them so valuable that they were often traded as currency. Abandoned homes and older buildings were even burned to the ground just to collect the nails.

- An average wood-frame home today can require over 20,000 nails to build!

The Wheel:

It's hard to say who made the first wheel, but it's over 5,000 years old. That may seem like a long time ago, but the wheel is actually a relative latecomer in early human inventions, with cloth, shoes, rope, glue, ships, boomerangs, simple plumbing, and the flute arriving many years earlier. Even the first wheels, originating in Mesopotamia, were not used for transport, but rather to spin clay for beautiful pottery. It took another 300 years before wheels were used for chariots, and the earliest sign of the Greeks possibly using wheelbarrows to carry loads at construction sites dates around 600 BC. Now we use wheels daily for travel, pulley systems, mechanical gears, and so much more. So the next time you ride in a car or pull something heavy in a wagon, thank these ancient civilizations for getting the wheel rolling.

Inventor tip:
Finding creative ways to do hard tasks can make them easier.

Fun facts:

- The Aztecs used wheels on toy figurines but not for transportation, probably due to unsuitable landscapes and a lack of large animals to pull carts.

- The wheel is sometimes called a truly human invention because, apart from some organisms that use rolling to their advantage, there are no clear models of the wheel in nature.

- The Ferris wheel, designed to compete with the wonder of the Eiffel Tower, was the centerpiece of the 1893 World's Columbian Exposition in Chicago. It was 76.2 meters (250 feet) in diameter and could carry over 2,000 people!

- The wheel has also had a big impact on language, religion, and philosophy. For example, its circular movement has been used as a symbol of rebirth in many cultures.

5

The Compass:

Early navigators often used the stars and physical landmarks for direction. This presented challenges when exploring far from the shorelines on the open sea or on cloudy nights when travelers could not always rely on stars to point the way. The magnetic compass immediately improved the way we explored the earth. The Chinese designed an early model of the compass during the Han Dynasty (206 BC–AD 220) using lodestone, which has a natural charge that aligns with the earth's magnetic field. The lodestone was first suspended by a rope so it could move freely toward the north-south direction. They later crafted a spoon-shaped piece of lodestone called the South Pointing Fish that was placed on a floating piece of wood. Over time, a variety of magnetized metals began to be used across the globe as needles that pointed north. The compass continues to be improved with new technology, but the natural laws that make it work remain constant, and sea captains and day hikers alike still depend on it to find their way home.

- Advanced GPS technology has increased our ability to find locations, but compasses are still used in planes, ships, and other vehicles because they work even when electronic signals are blocked.

- Compasses were disguised as ordinary objects and smuggled into internment camps during World War II to help prisoners escape and find their way to safety.

- Orienteering is the name of a sport based on direction finding. Competitors race to find checkpoints on a set course using a detailed map, and they rely on a thumb compass to keep the map facing the right direction.

- If you stand at the magnetic north pole, then your compass needle will spin in circles!

Inventor tip:
Learning about God's natural laws can lead to exciting new discoveries.

The Printing Press:

Stop the presses! This next invention actually created the front-page news. By inventing the Gutenberg Press (designed between 1436-1450), Johannes Gutenberg changed the way the world shares knowledge, but the story begins long before that. As early as AD 868, the Chinese developed block printing by carving letters in reverse into wooden boards to press ink against various types of paper. By the twelfth century, printed books were common among the wealthy in China, but it was the Gutenberg Press that sparked a printing revolution. Gutenberg used molten lead to design moveable letter pieces that could be adjusted within a machine to speed up the printing process. As the Gutenberg Press traveled throughout Europe, it increased literacy rates, spread news, advanced scientific discovery, and challenged systems of power by placing books and pamphlets in the hands of common people. Today, we can share our thoughts electronically with a simple click, but that would not be possible without the influence of the printing press.

Inventor tip:
Sharing ideas can speed up the progress of inventions.

Fun facts:

- Even in the digital age, we still use the term typesetting to describe the layout of books in recognition of the process of setting letters in place for the printing press.

- Johannes Gutenberg produced the now-famous Gutenberg Bible around 1455 to give glory to God for his new invention.

- Martin Luther's German translation of the New Testament is sometimes called the first best-seller, with 5,000 copies sold about two weeks after printing.

- The first official newspaper, *Relation*, was printed in 1605 in Strasbourg, France. Due to the impact of the printing press on the spread of news, reporters are still referred to today as the Press.

The Automobile:

Buckle up! We are about to hit the open road with the automobile. The idea of a mechanized carriage appeared in writing as early as Homer's *Iliad* (around 800 BC). The actual development of the automobile, however, came in stages. The steam engine inspired the French inventor Nicolas Cugnot to create a slow-moving, steam-powered vehicle in 1769. Though there were many additional attempts at vehicle designs, Karl Benz is often recognized as the creator of the first modern automobile, which was patented in 1886. He used a gasoline-fueled engine to power a buggy with three wheels. Then, the world changed forever in 1908 when Henry Ford produced his famous Model T. Using assembly lines, specialized labor, and interchangeable parts, Ford built cars in large quantities for a lower price. Originally marketed as a farm vehicle, the Model T quickly became popular with the general public, providing new opportunities for leisure, travel, and trade. Can you imagine a world without cars? Perhaps that is why many have called the automobile one of the most impactful inventions.

Fun facts:

- Leonardo da Vinci sketched a complex design around 1478 for a self-propelled cart using the force of springs and an advanced steering system.

- In 1903, Mary Anderson made traveling by car in a variety of weather conditions possible when she invented the windshield wiper.

- During World War II, automakers in the United States stopped making cars to produce military supplies to help end the war.

- Many "American" foods—such as hamburgers, fries, and shakes—became popular after new roadside diners were opened to feed hungry travelers on longer car trips.

Inventor tip:
Dividing big tasks into smaller jobs can make your work more efficient.

Railways:

All aboard! Next stop, the invention of the railway. Though ancient railway systems existed as early as 600 BC, it wasn't until the 1500s that wooden rails called wagonways first appeared in Germany. By 1776, iron tramways began appearing throughout Europe. Over time, the strength and durability of railways improved, but horses still pulled the wagons and carts. Then came the mighty steam engine. British inventor Richard Trevithick built the first steam-powered locomotive for tramways, but George Stephenson is considered the father of railways in England, as his lines were the first to be used for public transport. After Colonel John Stevens constructed the first locomotive in the United States in 1825, many new railways were built with large train companies powering the future of travel and trade. The most ambitious rail project was the Transcontinental Railroad, which connected the Central Pacific line starting in California to the Union Pacific railroad in Nebraska. A golden spike was placed to complete the project at Promontory, Utah, on May 10, 1869. Though train usage has been in decline since the 1920s, there is no doubt that railways set the world on a new track.

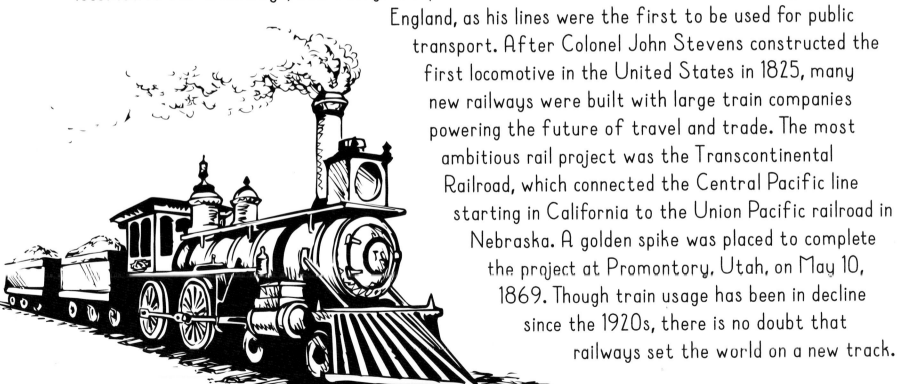

Fun facts:

- The ancient Greeks carved grooves into paved roads to make transport easier. The Diolkos, which helped carry boats by land over the Isthmus of Corinth, is one of the most famous examples.

- Many brilliant women made major railway improvements. Some examples include better axle bearings that decreased derailments (Eliza Murfey, 1870), a sound-dampening system to reduce noise (Mary Walton, 1881), and railway crossing gates for safety (Mary I. Riggin, 1890).

- The Czech Republic has more railways per square mile than any other country.

- High-speed trains traveling at 321.87 kilometers per hour (200 miles per hour) are now found throughout the world. New experimental trains have achieved test speeds close to 1,126.54 kilometers per hour (700 miles per hour)!

Inventor tip:
Drafting plans first can help your project not get derailed in later stages.

Inventor tip:
Dreaming big means the sky's the limit for your ideas.

The Airplane:

Ready to take flight? From ancient legends, like the story of Icarus, to space voyages, flight has always filled the human spirit with wonder. The first attempts to fly were modeled after nature. Inventors from across the globe attempted to mimic the flight of birds using gliders or movable wings. By the 1780s, other inventors designed lighter-than-air floating devices, such as baskets with balloons filled with hydrogen or hot air, but these were not easily controllable. Starting in 1799, Sir George Cayley made some important discoveries. He studied the forces of lift and drag to propose a new heavier-than-air concept for flying machines with fixed wings.

Many inventors built upon the work of Cayley, including Orville and Wilbur Wright who made the first controlled flight with a heavier-than-air plane at Kitty Hawk, North Carolina, on December 17, 1903. The first flight by Orville lasted twelve seconds, with a distance of 36.58 meters (120 feet). The triumph of the Wright brothers led to other successful flights in countries throughout the world. By 1927, Charles Lindbergh made the first non-stop, solo flight across the Atlantic Ocean, and only 42 years later, Apollo 11 landed on the moon! The dream of human flight is definitely a reality now, as high-tech aircraft are used for shipping, the military, travel, and adventure around the globe and beyond.

Fun facts:

- The original Wright brothers' airplane is preserved on display at the National Air and Space Museum in Washington, DC.

- The United States government purchased its first military plane from the Wright brothers in 1909. The price was $25,000, but the Wright brothers received a $5,000 bonus payment because their plane reached a flight speed of over 64.37 kilometers per hour (40 miles per hour).

- Amelia Earhart's historic flight across the Atlantic Ocean inspired other women to also push the limits of flight. For example, Jean Batten of New Zealand became famous in the 1930s for her record-breaking flights.

- In 2018, Virgin Galactic launched a new flight industry called space tourism that takes civilian passengers on trips into space.

The Telephone:

Have you ever tied two cans together to talk with a friend? This is similar to Robert Hooke's 1667 design of the acoustic telephone. His work showed that sound waves could travel greater distances when vibrating along a tight wire. The next major step forward was the telegraph, which could transmit messages using electricity, but it was limited to only one signal at a time.

There is debate over who made the first leap from the telegraph to the telephone. Antonio Meucci, Alexander Graham Bell, and Elisha Gray each designed a voice communication device in the 1870s, but after legal debates over patents, the US courts awarded Bell the right to market his new technology. By 1878, the first commercial telephone lines were established, and only three years later, almost 49,000 telephones were connected. In 1927, the first transatlantic call took place between the United States and England. Satellite technology then led to the development of the first cellular phone tests in the 1970s and the launch of the first digital network in Orlando, Florida, in 1993. Now, the world can connect instantly through smartphones that allow us to talk with family, send text messages, video chat, search the internet, and so much more.

10:10
Saturday, November 10
100%
swipe up to unlock

Inventor tip:
Healthy competition can motivate you to produce your best work.

Fun facts:

- Alexander Graham Bell's success with the telephone was inspired by his love for music. He knew multiple sounds could travel down a wire at the same time as long as the "notes" used different pitches.

- Born of enslaved parents, Lewis Latimer defied the odds to become an influential inventor. He was a self-taught draftsman and a patent expert who created the actual drawings of the telephone for Bell's patent application.

- On September 1, 1878, Emma Nutt made history as the first female telephone operator. As women occupied this important role with accuracy and diplomacy, they paved the way for more employment opportunities and better working conditions for other women.

- In 1929, Herbert Hoover became the first US president to use a telephone directly at his White House desk.

The Lightbulb:

Thomas Edison is often cited as the inventor of the lightbulb, but many others also contributed their bright ideas years before. As early as 1761, Ebenezer Kinnersley experimented with heating wires as a light source, but it was Humphry Davy in 1802 who lit the way for other inventors by passing a current through a strip of platinum to produce an early model for electric light. In 1879, English inventor Joseph Swan illuminated Mosley Street in Newcastle upon Tyne in a public demonstration of his new arc lamp, which some consider the first modern lightbulb. He replaced glowing platinum strips with carbonized paper filaments, which produced a strong glow, but burned out too quickly. Edison and his team worked tirelessly to solve this issue. They tested thousands of materials before discovering that a thin, carbonized, bamboo filament could burn for much longer. From there, the lightbulb has continued to be improved, with LEDs (light-emitting diodes) currently being the most popular bulbs in the world due to their energy efficiency and long life. From nightlights to computers, automobile headlights to keychain flashlights, the lightbulb has brightened the way we work, travel, and play.

Inventor tip:
Patiently improving your work over time can lead to better outcomes.

Fun facts:

- Swan and Edison had intense debates in court over who had the rights to sell the lightbulb. In the end, they joined their companies together to form the Edison-Swan United Electric Company.

- In 1881, Savoy Theatre in London was the first public building to be lit with electric lamps.

- Thomas Edison became so famous for the public display of his best inventions at Menlo Park, a small community in New Jersey, that he was known as the Wizard of Menlo Park.

- The latest LED lightbulbs can now last up to 50,000 hours!

The Sewing Machine:

Though the sewing machine is not always recognized as one of the most noteworthy human inventions, it definitely had a major impact on the world. In 1755, the German inventor Charles Wiesenthal submitted a patent in England for a needle designed for mechanical sewing. We don't know if he ever created the actual sewing machine he envisioned, but it inspired many other inventors to develop sewing machine prototypes throughout the early 1800s. By 1830, Barthélemy Thimonnier had started a sewing machine business to make uniforms for the French army. Unfortunately, the project ended in disaster when a mob of French tailors burned down his factory, fearing that their own hand sewing would be replaced by machines. By 1845, Elias Howe made more advances in sewing machine designs, but his machines were too expensive for families to buy. Taking many of these ideas, Isaac Singer then changed the sewing world. He made beautiful sewing machines at a lower cost, which lightened the burdens of mothers who often spent long hours stitching clothes by hand. Many businesses also boomed with the sewing machine. Today, the sewing machine continues to stitch together the fabrics of industry and creativity in important ways.

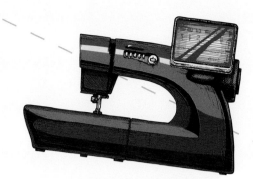

Inventor tip:
Focusing on helping others can expand the impact of your ideas.

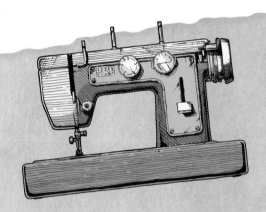

Fun facts:

- In 1845, Elias Howe had a competition where his sewing machine beat five humans with a top speed of 250 stitches per minute.

- Isaac Singer not only improved the sewing machine, but he also invented the idea of installment plans that allowed people to make payments for his products over time.

- Mahatma Gandhi, who sometimes criticized the impact of modern technology on the world, admitted that the sewing machine was "one of the few useful things ever invented."

- The Wright brothers depended upon a Singer sewing machine to make the coverings for their airplane wings.

21

The Battery:

Did you know the idea for the modern battery came from a frog? Well, not exactly, but when Italian biologist Luigi Galvani observed in 1780 that a dead frog's leg twitched when situated between brass hooks and his iron dissecting scalpel, he concluded that there was "animal electricity" remaining in the leg. His friend and scientist colleague, Alessandro Volta, was intrigued by the concept but disagreed with Galvani's theory. He conducted studies in 1791 that showed that differing metals with a wet conductor between them could create an electric current. By 1800, Volta developed the first battery with continuous current. He called it the Voltaic Pile because it was originally a stack of copper and zinc disks with soaked cardboard or cloth between them. Over the next century, many other inventors improved the battery. Gaston Planté, for example, invented the first rechargeable, lead-acid battery in 1859. Alkaline batteries started to be developed around the dawn of the twentieth century. They are still the most common battery type, but now lithium batteries are used in devices like cell phones, computers, and tablets. Today, batteries come in many shapes and sizes, but they all convert chemical reactions into electricity to power our world.

Fun facts:

- Benjamin Franklin coined terms such as "battery," "charge," "positive," and "negative" around 1749 when he was conducting his famous experiments on electricity.

- In 1936 small clay jars with an iron rod and copper coverings were discovered in Iraq. Made around 200 BC, the "Baghdad Battery" might be the oldest battery model ever created.

- Today, we still measure electric power in volts or voltage, named in honor of Alessandro Volta.

- Over 10 billion alkaline batteries are produced every year!

Inventor tip:
Being open to correction can further your search for truth.

The Refrigerator:

Preserving food has always been a high priority for humans. Before the refrigerator, packed ice or underground cellars extended the life of perishable foods. Artificial refrigeration, which typically removes heat from an enclosed space through evaporation, was first displayed publicly by Scottish professor William Cullen in 1756. His ideas, however, were not put into practice until years later. In 1805, American inventor Oliver Evans designed a machine that used vapor rather than liquid for cooling, but it wasn't until 1834 that Jacob Perkins actually made the first working refrigerator of this type. The electric refrigerator was first sold commercially by Fred W. Wolf in 1913. A year later, Florence Parpart patented an attachment that used electricity to circulate water through the refrigerator to keep it colder. She sold her system at a high price to large companies, but William C. Durant's launch of the Frigidaire Company truly grew the refrigerator market. Unfortunately, the expanded home use of refrigerators also led to problems, including deaths caused by toxic gas leaks. Freon then replaced ammonia, methyl chloride, and sulfur dioxide as the standard coolant in refrigerators.

Environmental concerns over Freon have prompted researchers today to search for even safer ways to keep food cold, including the use of solar and magnetic energy. Thanks to the work of many talented inventors, the refrigerator is truly one of the coolest inventions to date.

- Thomas Jefferson took pride in using his ice houses to make frozen treats for guests. Though he did not invent ice cream, he recorded the first known written recipe in the United States.

- Lillian Gilbreth, the first woman admitted into the Society for Industrial Engineers, was the inventor of shelving on refrigerator doors.

- "Smart" refrigerators are now on the market with internet, cameras, touch screens, and televisions.

- November 15th is celebrated as Clean Out Your Refrigerator Day.

Inventor tip:
Always take safety precautions when trying out new inventions.

 # The Television:

Though dreams of transporting pictures over distances started long before, the mechanical television came into focus when Paul Nipkow made spinning disks with holes placed in spiral formations to produce individual lines of light that created images. Many viewed the Nipkow Disk (1884) as the future of television. In the early 1900s, however, there was a split among inventors after German physicist Karl Braun developed technology that used electricity in vacuum tubes instead of spinning disks. The first electric television set was ultimately created by a farm boy who lived in a small log cabin in Utah without electricity for most of his early life. During his high school years, Philo Farnsworth envisioned an improved way to code radio waves in a vacuum tube to form moving images that would be projected on a screen. At age 21, on September 7, 1927, in San Francisco, California, he made that vision a reality by transmitting a single line.

Major broadcasting companies such as RCA took an active interest in this new media, and by the 1950s, television started having a major impact on advertising, politics, and the value systems of society. The internet further transformed programming with on-demand streaming and instant access to the latest news and entertainment. The television has certainly come a long way from its idea stages on a potato farm to the global impact it has today!

Fun facts:

- Though considered by many as the father of the television, Philo Farnsworth did not want his own children to watch it, as he did not think it was of intellectual value.

- Television usage has been in decline since 2016 due to the rise in popularity of social media.

- Many political experts believe that John F. Kennedy beat Richard Nixon in the US presidential election of 1960 because he looked more youthful and energetic on the new media of television.

- Over a lifetime the average person will view over 2 million television commercials.

The Dishwasher:

Though Joel Houghton was the first to patent the dishwasher in 1850, his wooden box with a hand crank did not prove to be an effective design. It would be over 30 years before another resolute inventor would change kitchens forever. Josephine Cochrane was born into a family with a heritage of inventing. When her fine china got chipped by workers after a dinner party, she began to wash them carefully herself, but then realized there had to be a better way.

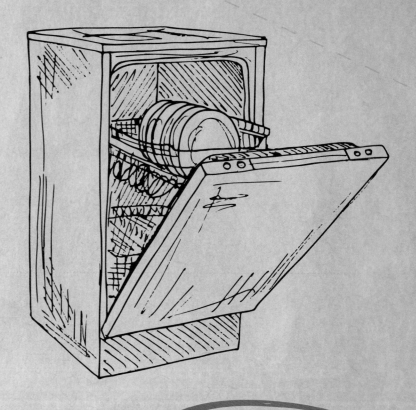

"If nobody else is going to invent a dishwashing machine, I'll do it myself."
—Josephine Cochrane

Inventor tip:
Keep working toward your dreams despite the challenges.

She put her creative mind to work and designed a dishwashing machine that used water pressure instead of brushes for cleaning. This would earn her a patent on December 28, 1886. Cochrane boldly promoted her dishwashing machine at the 1893 World's Columbian Exposition in Chicago. She worked hard to support her family by selling units to hotels and restaurants. By 1924, other inventors, such as William Howard Livens, made the dishwasher more suitable for personal use, and a few decades later, the dishwasher became a common appliance. Today, not all homes have automatic dishwashers, and some families choose to wash their dishes by hand. Regardless of your own method of cleaning dishes, Josephine Cochrane's story provides an example of perseverance no matter what life dishes out.

Fun facts:

- Josephine Cochrane called her washing machine Lavaplatos, which is the word for dishwasher in Spanish.

- A high-efficiency dishwasher can save a household thousands of gallons of water each year over handwashing.

- For most dishwashers today, prewashing dishes actually makes them less effective, especially since detergents are designed to adhere to food particles.

- Running your dishwasher at night can save on electricity bills, since in many cities it is not a peak energy hour and electricity use costs less.

The Computer

The original word "computer" actually referred to an occupation and not a machine. Computers were mostly mathematicians and clerks who were employed to solve complex number problems. Inventors in the 1800s, such as Charles Babbage, wanted to develop machines that could "compute" large number sets faster than humans. Ada Lovelace, the daughter of the poet Lord Byron, published notes on Babbage's inventions that are recognized by many as the first computer programs. It wasn't until the 1930s that progress toward the modern computer accelerated with many bright minds contributing new ideas. The computer microchip was developed in 1958 by Jack Kilby and further refined by Robert Noyce the following year.

Starting in the 1970s, familiar names such as Bill Gates and Steve Jobs were working hard to figure out ways to take the computer into people's homes. Apart from desktops, laptops, and phones, computer technology is now found in everything from toys to cars to microwaves. Far beyond their original purpose to calculate numbers, computers have completely transformed many aspects of society and our personal lives.

Fun facts:

- One of the most famous early computers, called the ENIAC, was enormous. It weighed over 27 tons and required around 167.23 square meters (1,800 square feet) of space.

- Computer programming pioneer Grace Hopper and her team at Harvard University coined the terms "bug" and "debug" when a large moth got into their computer and stopped it from working.

- Three of the largest computer companies in the world—Apple, HP, and Microsoft—were each started in a garage.

- If the human brain were measured in computer terms, its memory capacity is estimated by some researchers to be as large as 2,500 terabytes!

Inventor tip:

Take time to recognize the hand of God as you develop your ideas.

The stories in this book represent only a handful of the amazing inventions that God has blessed us to receive. Each one presents an extraordinary triumph, but remember that the most important chapters in the story of invention are yet to be written. What ideas do you have? As you review the inventor tips below, remember that you have the power to create things that can make the world a better place!

An idea is never too small to have a big impact.

Drafting plans first can help your project not get derailed in later stages.

Being open to correction can further your search for truth.

Finding creative ways to do hard tasks can make them easier.

Dreaming big means the sky's the limit for your ideas.

Always take safety precautions when trying out new inventions.

Learning about God's natural laws can lead to exciting new discoveries.

Healthy competition can motivate you to produce your best work.

You are never too young to have world-changing ideas.

Sharing ideas can speed up the progress of inventions.

Patiently improving your work over time can lead to better outcomes.

Keep working toward your dreams despite the challenges.

Dividing big tasks into smaller jobs can make your work more efficient.

Focusing on helping others can expand the impact of your ideas.

Take time to recognize the hand of God as you develop your ideas.